The Moon

by Joanne Mattern

FOCUS READERS®

BEACON

www.focusreaders.com

Focus Readers is distributed by North Star Editions:
sales@northstareditions.com | 888-417-0195

Produced for Focus Readers by Red Line Editorial.

Photographs ©: Goddard Space Flight Center/Arizona State University/NASA, cover, 1; Shutterstock Images, 4, 8, 12, 14, 18, 20–21, 22, 25; JSC/NASA, 6, 17, 29; MSFC/NASA, 11; Joel Kowsky/HQ/NASA, 26

Library of Congress Cataloging-in-Publication Data
Names: Mattern, Joanne, 1963- author.
Title: The moon / by Joanne Mattern.
Description: Lake Elmo, MN : Focus Readers, [2023] | Series: Space | Includes index. | Audience: Grades 2-3
Identifiers: LCCN 2022006598 (print) | LCCN 2022006599 (ebook) | ISBN 9781637392461 (hardcover) | ISBN 9781637392980 (paperback) | ISBN 9781637394007 (pdf) | ISBN 9781637393505 (ebook)
Subjects: LCSH: Moon--Juvenile literature.
Classification: LCC QB582 .M37 2023 (print) | LCC QB582 (ebook) | DDC 523.3--dc23/eng20220405
LC record available at https://lccn.loc.gov/2022006598
LC ebook record available at https://lccn.loc.gov/2022006599

Printed in the United States of America
Mankato, MN
082022

About the Author

Joanne Mattern has written many nonfiction books for children. Her favorite topics are science, sports, history, and animals. Joanne enjoys showing young readers the fascinating people, places, and things in our world. She lives in New York State with her family. She enjoys doing jigsaw puzzles and taking long walks.

Table of Contents

Chapter 1

Moon Landing

On July 16, 1969, a spacecraft blasted off from Florida. Three astronauts inside were flying to the Moon. Two of them would be the first to walk on the Moon's surface.

The astronauts lifted off in a Saturn V rocket.

The astronauts landed in an area called the Sea of Tranquility. They planted a US flag there.

They traveled for three days. Then their spacecraft began **orbiting** the Moon. The next day, two of the astronauts landed on the

Moon's surface. One stayed in the spacecraft above the Moon.

The two astronauts collected Moon rocks. They took photos. They planted a US flag. Then they went back to their lander. They tried to sleep. Later, all three astronauts headed home to Earth together.

The spacecraft had several parts. The lander split off to land on the Moon's surface. The rest stayed in orbit.

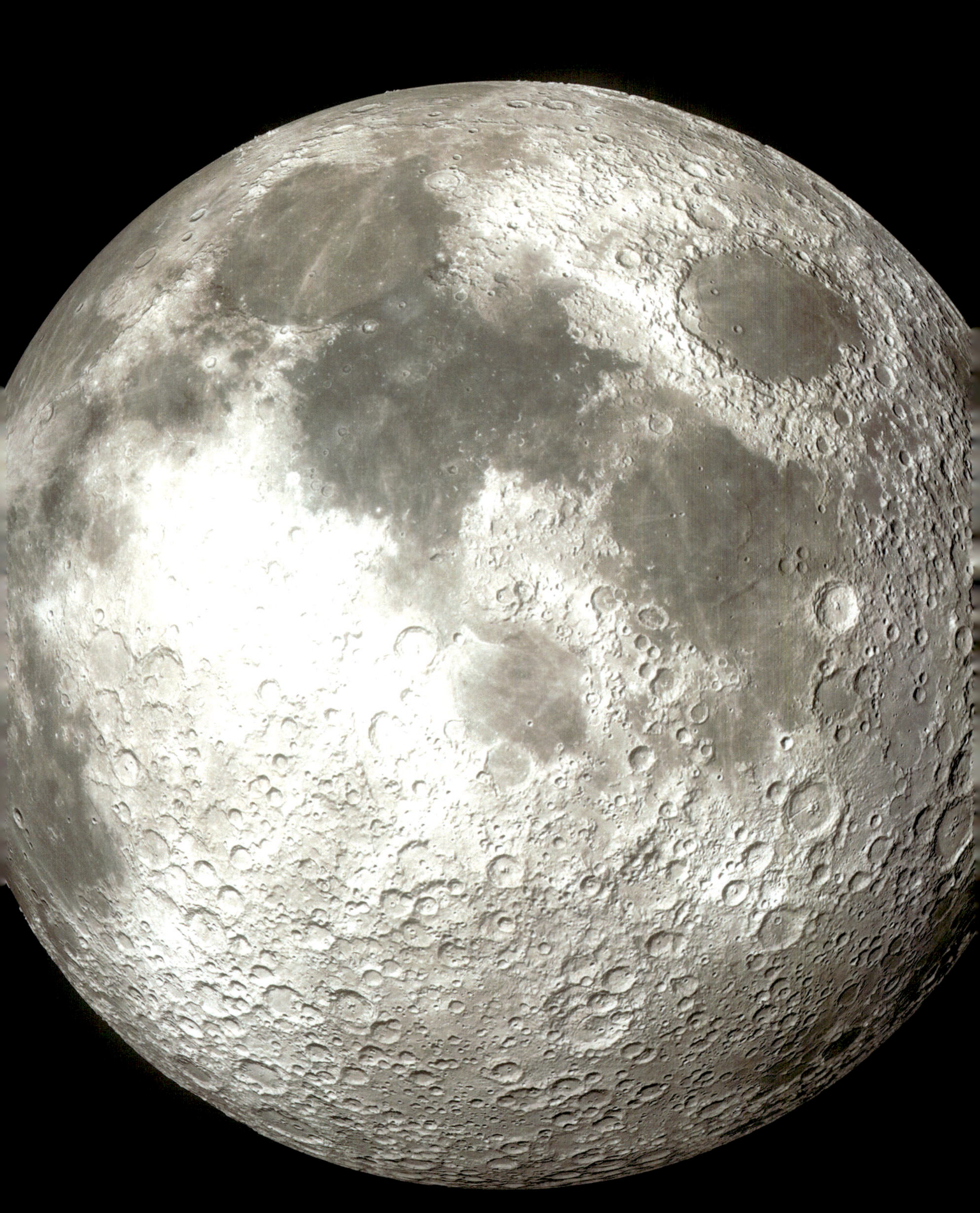

Chapter 2

Our Closest Neighbor

The Moon is billions of years old. Scientists believe the Moon was formed when a huge object hit Earth. The object was as big as Mars. When it hit Earth, a large chunk of rock broke off the object.

The Moon is approximately one-fourth as wide as Earth.

That rock began orbiting Earth. It became Earth's Moon.

The Moon has three parts. One part is deep inside the Moon. It is called the core. The core is made of iron and **molten** rock.

The middle layer of the Moon is the mantle. The mantle is made of iron and other minerals. The outer surface of the Moon is called the crust. The crust is made of iron, aluminum, and many other minerals. The crust is 43 miles

The Moon's surface is covered in rocks and dust.

(70 km) thick on one side of the Moon. It is 93 miles (150 km) thick on the other side.

Objects in space often hit the Moon. These objects leave craters.

Compared to Earth, the Moon has very little water.

A crater is a bowl-shaped hole. The Moon's surface has thousands of craters. Tycho Crater is one of the largest. It is more than 52 miles (85 km) wide. We can see the Moon's largest craters all the way from Earth. The Moon is approximately 238,900 miles (384,400 km) away.

The surface of the Moon is also covered with rocks, dust, and soil. Some parts look dark. Others look light. The light parts were once filled with lava.

The Moon is a dry place. But probes have discovered water under the surface.

Chapter 3

Round and Round

The Moon orbits Earth. It takes approximately 29 days. The Moon also rotates. It turns at the same rate that it circles Earth. So, the same side of the Moon always faces Earth.

The Moon rises in the sky approximately 50 minutes later each day.

The Sun shines on the Moon. However, the Moon is always moving. For this reason, we do not always see all of the Moon. Instead, we see the Moon in **phases**. The Moon has four main phases. Each phase shows a different amount of the Moon.

A full moon happens when the Sun lights up the Moon’s whole face. During a full moon, the Moon looks big and bright in the night sky.

The Moon is truly full for just a moment. But from Earth, it looks full for two or three nights.

During a first quarter moon, we can see part of the Moon. The amount we see gets smaller every night. Soon, we cannot see it at all.

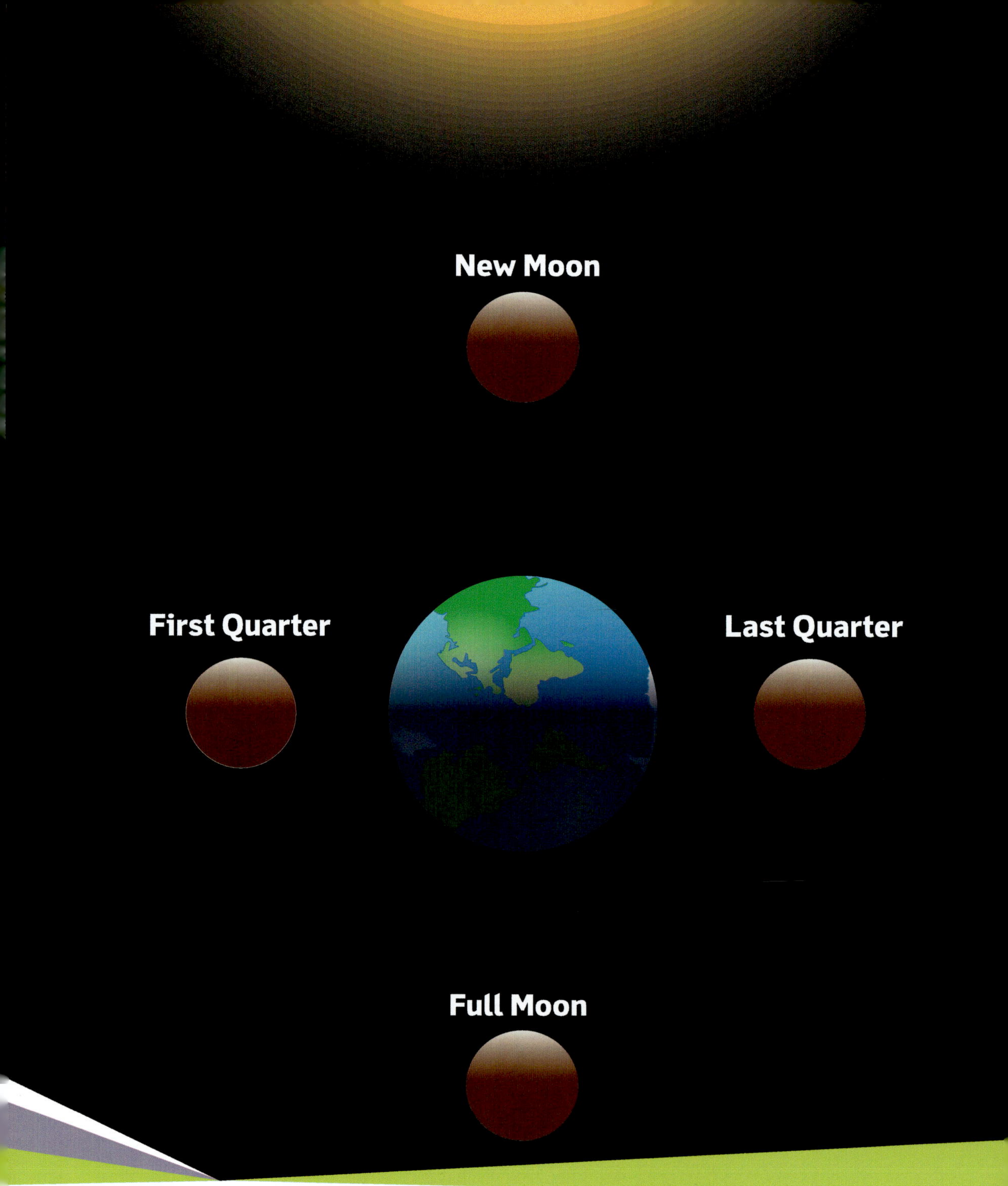

New Moon

First Quarter

Last Quarter

Full Moon

The Moon has four main phases. In between, it goes through other phases.

This is called a new moon. During a new moon, the bright side faces away from Earth.

After the new moon, the Moon starts to look bigger. This is the last quarter moon. In about a week, the Moon will be full again.

Did You Know?

Sometimes, Earth comes between the Moon and the Sun. Earth's shadow falls on the Moon. This event is called a lunar eclipse.

THAT'S AMAZING!

A Helpful Neighbor

Life on Earth would be very different without the Moon. The Moon's **gravity** pulls on Earth. This helps Earth stay steady. Without the Moon's pull, Earth would wobble. That wobble would make our **climate** unstable. Without the Moon, Earth could even tip sideways.

The Moon's pull also creates the **tides** on Earth. Also, many animals use the Moon. It helps them find their way.

Some birds use the Moon for migration. They fly south in winter soon after the full moon.

Chapter 4

Exploring the Moon

The first trip to the Moon took place in 1969. US astronauts landed on the Moon five more times after that. These **missions** helped scientists learn more about the Moon.

The first mission where people landed on the Moon was called Apollo 11.

They learned how old the Moon is. They also learned what the universe was like billions of years ago.

After 1972, US astronauts stopped traveling to the Moon. The cost of going to the Moon was very high. People wanted to spend that money in other ways.

Other countries have explored the Moon, too. Russia, Japan, and India all sent spacecraft to study the Moon. Between 2013 and 2020, China landed **rovers** on the Moon.

Rovers can send information back to Earth. They do not need people to collect information.

The rovers took many photos. They also brought samples from the Moon back to Earth.

In 2021, the United States began preparing for new missions.

New space suits were designed for the Artemis program.

A new US program planned to send astronauts to the Moon again. This program was called Artemis. Japan also planned to send a rover to the Moon. Scientists were excited to learn more about Earth's closest neighbor.

Did You Know?

The Moon is slowly moving away from Earth. It moves 1.48 inches (3.78 cm) away every year.

FOCUS ON
The Moon

Write your answers on a separate piece of paper.

1. What is the main idea of Chapter 2?
2. Do you think it is important to explore the Moon? Why or why not?
3. During which phase can we see the whole face of the Moon?
   - A. full moon
   - B. new moon
   - C. quarter moon
4. Why did US astronauts stop going to the Moon?
   - A. Astronauts wanted to go to Mars instead.
   - B. Scientists no longer wanted to learn about the Moon.
   - C. Trips to the Moon cost more than people wanted to spend.

5. What does **astronauts** mean in this book?

*On July 16, 1969, a spacecraft blasted off from Florida. Three **astronauts** inside were flying to the Moon.*

- **A.** people who study engines
- **B.** people who build rockets
- **C.** people who go into space

6. What does **rotates** mean in this book?

*The Moon also **rotates**. It turns at the same rate that it circles Earth.*

- **A.** spins around
- **B.** falls apart
- **C.** stops moving

Answer key on page 32.

Glossary

climate
The average weather conditions of a particular place or region.

gravity
A force that pulls objects toward one another.

missions
Important tasks or operations.

molten
Melted by intense heat.

orbiting
Repeatedly following a curved path around another object because of gravity.

phases
Changes in how the Moon looks in the sky as it orbits Earth.

probes
Devices used to explore.

rovers
Wheeled spacecraft that roll across the surface of a planet or moon.

tides
Daily rises and falls in sea level.

To Learn More

BOOKS

Read, John A. *50 Things to See on the Moon: A First-Time Stargazer's Guide*. Halifax, NS: Formac Publishing Company Limited, 2019.

Ringstad, Arnold. *The Moon*. North Mankato, MN: The Child's World, 2021.

Tomecek, Steve. *The Moon*. New York: Children's Press, 2021.

NOTE TO EDUCATORS

Visit **www.focusreaders.com** to find lesson plans, activities, links, and other resources related to this title.

Index

Answer Key: 1. Answers will vary; **2.** Answers will vary; **3.** A; **4.** C; **5.** C; **6.** A